Apples  for Teachers

A Basic Skills Reinforcement Program for Young Children

Time:
Months, Holidays, Seasons, and Telling Time

by
Diane Burkle
Cynthia Polk Muller
and
Linda J. Petuch

Fearon Teacher Aids
Carthage, Illinois

contents

Introduction 3
Using the Worksheets 4
Setting Up Learning Centers 12
Worksheets
Months and Holidays
    September 13
    October 18
    November 23
    December 29
    January 35
    February 40
    March 45
    April 50
    May 55
    June 60
    July 65
    August 70
Seasons
    Fall 75
    Winter 78
    Spring 81
    Summer 84
    Reviewing the Seasons 87
Telling Time 89

Illustrator: Marilynn G. Barr

Entire contents copyright © 1988 by Fearon Teacher Aids, 1204 Buchanan Street, P.O. Box 280, Carthage, Illinois 62321. However, the individual purchaser may reproduce designated materials in this book for classroom and individual use, but the purchase of this book does not entitle reproduction of any part for an entire school, district, or system. Such use is strictly prohibited.

ISBN 0-8224-0458-3

Printed in the United States of America

1. 9 8 7 6 5

introduction

Time: Months, Holidays, Seasons, and Telling Time is designed to enhance primary education through the use of manipulatives and tactile materials. The learning activities help students develop visual, auditory, kinesthetic, tactile, and perceptual skills while providing opportunities for creative expression. Although students may appear to be participating in art projects, they are, in fact, achieving artistically such curriculum objectives as recognizing the names of the months and seasons, sequencing numbers on a clock, distinguishing day from night, and developing eye-hand coordination and fine motor skills.

The worksheets contained here are structured to allow repetition when learning new concepts—an approach that educators have noted is essential for internalizing concepts and skills. The learning materials are specifically designed to appeal to young children and to provide them with the positive reinforcement they need for learning new skills.

Worksheet-based activities require minimal teacher preparation and can be used with students working individually, in small groups, or in independent learning centers. Most of these activities use common school supplies such as crayons, scissors, and glue or paste. Some activities require additional materials that are inexpensive and readily available in many classrooms or from art supply stores.

To administer an activity, assemble and distribute the necessary materials. (Materials lists, along with suggestions for variations, extensions, and enrichment activities, are provided in the section called "Using the Worksheets.") Read the worksheet instructions aloud to the students and make sure they understand what to do. Provide help as needed (especially with the use of a hole punch), and always supervise the use of scissors and glue.

We hope you enjoy using this book and, most of all, that it makes learning fun for your students.

using the worksheets

There are five worksheet activities for each of the twelve months, three worksheets for each of the four seasons, and eight worksheets devoted to telling time. Review worksheets for reinforcing season discrimination are also included. After your students have completed one or two worksheets of a kind, they may be able to complete the others of that kind without instructions. Use the following notes to help you administer each kind of worksheet:

calendar
(pp. 13, 18, 23, 29, 35, 40, 45, 50, 55, 60, 65, and 70)

objectives and skills
Recognizing the name of the month
Recognizing the names for the days of the week
Sequencing numbers 1–31

materials
crayons

variations
✓ Provide a completed calendar for your students to use as a model.

✓ Your students may not be able to write the numbers. In that case, write the numbers in place before you copy the page. Then have the students trace over the numbers.

extension
✓ Have the students outline special days, such as holidays, in red crayon. You might also provide special stickers for them to indicate very special days, such as their own birthdays.

month or holiday art projects
(pp. 14, 19, 24–25, 30–31, 36, 41, 46, 51, 56, 61, 66, and 71)

objectives and skills
Recognizing months
Associating symbols with common holidays
Writing numbers and own name (September, January, June)
Developing eye-hand coordination
Developing fine motor skills

variations
✓ Some activities call for glue. For some of these activities, you may wish to have students use glue sticks instead of glue or paste. Glue sticks are available at most stationery stores or art supply stores, they are convenient, and they keep artwork (and children's hands!) neat and clean.

✓ **September Name Tags** (p. 14): If students cannot write their names on their name tags, you will have to do this for them.

✓ **December Gingerbread** (p. 30): Add a drop of oil of clove or oil of cinnamon to the cotton balls before the students fill their gingerbread figures with the cotton. Help the children staple the gingerbread figures together, or do this for them. Punch a hole in the top of each completed figure, and string a length of yarn through the hole. Hang the gingerbread figures throughout the classroom.

✓ **New Year's Celebration** (p. 36): If students cannot write the year, do this step for them before you copy the worksheet.

✓ **February Cherry Tree** (p. 41): Students may color the tree trunk and leaves with crayons instead of gluing the hole-punched circles along the outlines.

✓ **March Lamb** (p. 46): Students may continue gluing the curled pieces of yarn all over the lamb, if they wish.

✓ **July Picnic** (p. 66): Have the students classify the food items in various ways (meats, vegetables, fruits, sweets; main course, side dish, dessert; and so on) before they glue them inside the basket. Students may draw additional food items, or find pictures of additional food items in old magazines and newspapers to cut out and glue inside the basket.

✓ **August Beach Things** (p. 71): Students may color the beach sand instead of gluing on real sand, if you prefer.

extensions

✓ **September Name Tags** (p. 14): Pin the name tags on a clothesline strung across one area of the classroom. Each day as the students come to class, have them find their own name tags. This way, you will be able to keep attendance by noting which T-shirt name tags are left on the clothesline.

✓ **Thanksgiving Place Setting** (p. 24): Have the students cut out pictures of food from old magazines and newspapers. They can then glue the pictures on their plates.

enrichment

✓ As each new month is introduced, point to that month on a 12-month calendar to show students how the month is positioned in the year. This will give students a sense of the longer-term passage of time.

✓ Explain the significance of the major holidays.

✓ Read or tell stories of poems that have holiday themes. A children's librarian should be able to give you a listing of appropriate holiday stories.

materials (per student)

September Name Tags (p. 14)
crayons
scissors
safety pins

October Ghost (p. 19)
glue
dried beans

Thanksgiving Place Setting (pp. 24–25)
crayons
scissors
11" × 17" construction paper

materials (per student)
(continued)

December Gingerbread
(pp. 30–31)
crayons
scissors
stapler
cotton balls

New Year's Celebration (p. 36)
crayons
glue
glitter

February Cherry Tree (p. 41)
hole punch
green and brown
 construction paper
glue
2″ squares of red tissue paper

March Lamb (p. 46)
black and pink crayons
1½″ lengths of black yarn
glue

April Rains (p. 51)
crayons
scissors
8½″ × 11″ construction paper
 of any color
glue

Children Around the Maypole
(p. 56)
crayons
scissors
8½″ × 11″
 construction paper
glue
6 pieces of yarn in varying
 lengths, from 2″ to 5″

Father's Day Card (p. 61)
crayons
scissors
8½″ × 11″
 construction paper

July Picnic (p. 66)
crayons
scissors
glue

August Beach Things (p. 71)
crayons
glue
sand

mosaic projects
(pp. 15, 20, 26, 32, 37, 42, 47, 52, 57, 62, 67, and 72)

objectives and skills
Recognizing names of months
Associating symbols with common holidays
Developing direction-following skills
Developing fine motor skills

variations

✓ **Thanksgiving Turkey Mosaic** (p. 26): You will have to dye the raw rice for this activity beforehand. Soak the raw rice in diluted food coloring for a few minutes. Drain it on newspaper. If you prefer, have the students glue dried beans, seeds, peas, and rice or different natural colors; this way, you won't have to dye the rice.

✓ **Valentine's Day Mosaic** (p. 42): Have students glue the small heart onto the large heart. Then display all of these double hearts on a bulletin board. Or, have each student glue the hearts on the outside of a lunch bag. The students may use the decorated lunch bags to collect Valentine's Day messages from other students.

✓ **Mother's Day Mosaic** (p. 57): If your students are unable to write, you should write a brief message (such as "Happy Mother's Day") on the writing line before you copy the worksheet for the students. Then have the students trace the message.

✓ **Fourth of July Mosaic** (p. 67): Encourage students to form a circle with the group of 13 stars. When they glue the blue paper in a mosaic fashion over the area on the flag, they will be covering up the stars shown. Point out the circular formation before they glue their blue-paper mosaics.

extension
✓ **Valentine's Day Mosaic** (p. 42): Capable students may write Valentine's Day messages on the construction paper on which the two hearts are glued.

enrichment
✓ Discuss the significance of each symbol represented. For example, the symbol for September is the apple; explain that September is the month when most apple harvests occur.

✓ Discuss events that take place during a given month. Be sure to ask for suggestions from the students. For example, students may mention that September is usually back-to-school month.

✓ Point out that some symbols for a given month may be appropriate for other months as well. For example, most symbols of winter (snowflakes, warm mittens, snow-covered pine trees) may be appropriate for almost any month in winter.

materials (per student)

September Apple Mosaic (p. 15)
crayons
glue
2" squares of red tissue paper
scissors
blank paper

October Cat-and-Moon Mosaic (p. 20)
glue
2" squares of orange tissue paper
scissors
blank paper
black crayon
four 1" lengths of black yarn

Thanksgiving Turkey Mosaic (p. 26)
glue
small paper plate (about 8")
raw rice, colored with nontoxic food coloring
crayons
scissors

December Sleigh Mosaic (p. 32)
crayons
foil wrapping paper
scissors
glue

January Snowman Mosaic (p. 37)
crayons
white paper
small black hole-punched circles
small triangle cut from orange paper

Valentine's Day Mosaic (p. 42)
glue
scissors
2" squares of red tissue paper
2" squares of white tissue paper
8½" × 11" construction paper (preferably pink)

March Shamrock Mosaic (p. 47)
green construction paper
glue

Easter Egg Mosaic (p. 52)
crayons
sequins, dried peas and beans of different colors, small colored marshmallows, and other small decorations
glue
strip of green construction paper, 8½" long by 2½" wide
scissors

materials (per student)
(continued)

Mother's Day Mosaic (p. 57)
crayons
glue
2" circles of green
  tissue paper
scissors
construction paper in
  various bright colors
dried peas

June Playground Mosaic (p. 62)
blank paper
glue
2" squares of brown
  tissue paper
scissors
crayons

Fourth of July Mosaic (p. 67)
white and red crayons
blue construction paper
glue
13 small star stickers,
  preferably white

August School Box Mosaic
(p. 72)
construction paper in any color
glue
old magazines and newspapers
  showing school supplies
  (A school supply or
  stationery supply catalog
  is an excellent alternative
  source.)
scissors

lace-up projects
(pp. 16, 21, 27, 33, 38, 43, 48, 53, 58, 63, 68, and 73)

objectives and skills
Recognizing names of months
Developing fine motor skills
Developing eye-hand coordination

materials
crayons
glue
8½" × 11" tagboard
scissors
hole punch
36" shoelaces, or yarn with each end taped tightly

variations
✓ Allow students to lace the shape in a variety of ways, including over and under—lacing around the outside of the shape (in a whipstitch)—or in and out—lacing down through one hole and up through the next. The first method is generally easier for students.

✓ Prepare the lacing cards for students to use on their own: laminate a trimmed copy of the worksheets onto tagboard, and punch out the holes. This way, students will only have to perform the lacing part of the activity.

enrichment
✓ Ask students to describe each lace-up shape and to tell a story about it. Encourage them to use their imagination.

✓ Explain the significance of each object to the month it represents.

writing the month name
(pp. 17, 22, 28, 34, 39, 44, 49, 54, 59, 64, 69, and 74)

objectives and skills	Reinforcing the name of the month Differentiating colors Developing direction-following skills
materials	crayons
extension	✓ Provide writing paper so that students may practice writing the name of the month.
enrichment	✓ Give the ordinal number of a month (first, second, and so on) and ask a student to point to that month on a 12-month calendar, counting out the months. Try the reverse: give the name of a month and ask a student to give its ordinal number (e.g., January: the first month).

season circles (pp. 75, 78, 81, and 84)

objectives and skills	Tracing the name of the season Recognizing symbols that represent seasons Developing eye-hand coordination Developing fine motor skills
materials	hole punch construction paper in specified colors glue
extensions	✓ Have the students draw pictures of other objects that relate to the season. Gather the pictures into a class booklet (or have each student form her or his own booklet) on the seasons. ✓ Display the completed pictures on the bulletin board.
enrichment	✓ Discuss the symbol representing a season. Encourage the students to tell what they know about that object and how it relates to the season. ✓ Have the students name the seasons in order. ✓ Have the students identify months that are in each season. Capable students may learn the day that each season begins.

sponge painting (pp. 76, 79, 82, and 85)

objectives and skills	Reinforcing the names of the seasons Developing fine motor skills Developing direction-following skills
materials	crayons tempera paint water shallow plastic containers for paint sponges, cut into 1" cubes optional: aprons for students to wear

seasonal scenes (pp. 77, 80, 83, and 86)

objectives and skills
Developing left-to-right orientation
Developing fine motor skills
Developing direction-following skills

materials
crayons
water colors
water
shallow plastic containers for paint
paintbrushes

enrichment
✓ Have students tell about a given scene in their own words. They might simply describe what they see, or they might make up a story to accompany the scene. You may copy their stories or narratives onto another sheet of paper and attach it to the completed worksheet picture.

✓ Have the students relate their own experiences of a given season.

✓ Ask the students to name characteristics of a given season. List these on the chalkboard, and have the other students verify those characteristics.

reviewing the seasons (pp. 87–88)

objectives and skills
Distinguishing the seasons
Reading and writing the names of the seasons

materials crayons

clock art/ hidden time (pp. 89–90)

objectives and skills
Identifying standard and digital clocks
Developing fine motor skills

materials crayons

variation
✓ Introduce these worksheets by discussing clocks and watches. Have the students read the numbers around a clock face, and have them point to the short (or little) hand and the long hand of the clock. Ask if anyone knows which hand is the hour hand and which is the minute hand. Show students a digital clock, too, if you wish, to make sure they recognize that this is also a clock.

clock numbers (pp. 91–92)

objectives and skills	Sequencing numbers on a clock Reading numbers Writing numbers
materials	crayons
variation	✓ For some students, these worksheets will be too difficult. Assign the worksheet that has four clocks per page to those students who need a challenge.
extensions	✓ Ask the students to explain why they circled the items they did. Not all answers are right or wrong on these worksheets. For example, some students may have noticed that the moon is sometimes visible in the daytime sky. And some may not have seen lightning in the sky, so they might not realize what it is. ✓ Discuss what color the sky is during the day and at night. Talk about the colors during sunrise and sunset. Discuss the colors of various items in the sky (sun, clouds, moon, and so on). ✓ Discuss other items that may be visible in the sky at night or during the day.
extension	✓ Say a time to the hour, and have the students draw hands on a clock to show the time named.
enrichment	✓ Make a clock face on the chalkboard, but include only about six of the numbers. Have a student volunteer to write the other numbers in place. Erase several numbers and have another student fill in the missing numbers.

telling time (pp. 93–94)

objectives and skills	Reading numbers Writing numbers Associating clock numbers with time to the hour
materials	crayons
extensions	✓ Use a demonstration clock to show how the hands move and the hours pass. ✓ Discuss other events that take place during certain times of the day. For example, eight o'clock might be breakfast time, twelve o'clock might be lunchtime, and two o'clock might be nap time.

day and night (pp. 95–96)

objectives and skills	Distinguishing day from night Associating objects with day or night Differentiating dark and light
materials	crayons

setting up learning centers

The worksheet-based activities in this book may be adapted to a learning center environment. Most worksheets focus on a particular month or season, and the worksheets are usually repeated (sometimes with slight variation) for every month or season. Hence, even young children who cannot yet read may be able to work at a learning center, without teacher direction, completing worksheets on their own or in small groups. Here are some suggestions that may help you in setting up learning centers in your classroom.

1. Set up a separate learning center for each subject area. Identify the center by hanging a symbol, such as a month or season label, over the table set up for that learning center.

2. Gather materials that students will need for the activity. Place those materials in boxes or other appropriate containers, and label the containers with the word and a picture of the item contained. Store these containers in the learning center.

3. For activities requiring paint, glue, or paste, tape a protective covering over the learning center table. Keep the covering on the table as long as that learning center continues to be used.

4. Prepare samples of the activities. Post the samples on the bulletin board or in the learning center so students can refer to them.

5. Explain each activity before the students begin working on their own. You may need to review the directions on a daily basis.

6. If you have a few learning centers set up, students may work at the centers in rotation. Group students by ability, by compatibility, or in some random fashion. Create a chart that shows who is in each group, and post the chart on a bulletin board. Create tags that match the learning center symbols (see suggestion 1), and pin a different tag next to each group listed on the chart. The tag will tell the group which center they should go to that day. Change the tags each day.

   Some activities are especially appropriate for cooperative learning situations. You may wish to assign roles, such as gathering materials, cleaning up after the activity, collecting the worksheets and making sure group members have written their names on their papers, and so on. You might also want to provide an appropriate social goal such as working quietly, asking group members for help, and so on.

7. Provide adequate space for the completed worksheets to dry, if they have just been glued, pasted, or painted. Try not to stack or overlap the worksheets while they dry.

September Calendar

Name _____

Number the days.
Color the picture below.

September						
Sunday	Monday	Tuesday	Wednesday	Thursday	Friday	Saturday

Time: Months, Holidays, Seasons, and Telling Time © 1988

Name _____

September Name Tags

Write your name on the name line below.
Color the T-shirt.
Cut out the T-shirt.

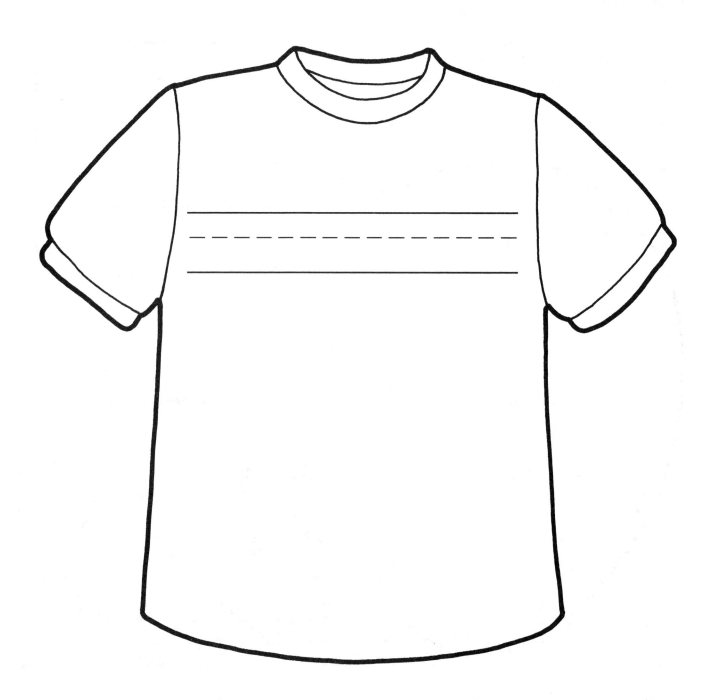

14

Time: Months, Holidays, Seasons, and Telling Time © 1988

September Apple Mosaic

Name _____

Color the leaves and the stem. Put glue on the apple. Place small pieces of red tissue paper over the glue, overlapping the pieces. Let the glue dry.
Then cut out the apple shape. Glue the apple to another piece of paper. Cut out the leaves and stem and glue them in place on the apple.

Time: Months, Holidays, Seasons, and Telling Time © 1988

Back-to-School Lace-up

Name _____

Color the schoolhouse.
Glue this page onto tagboard.
Cut out the schoolhouse.
Punch out a hole at each dot.
Sew around the schoolhouse with a lace, and tie the ends.

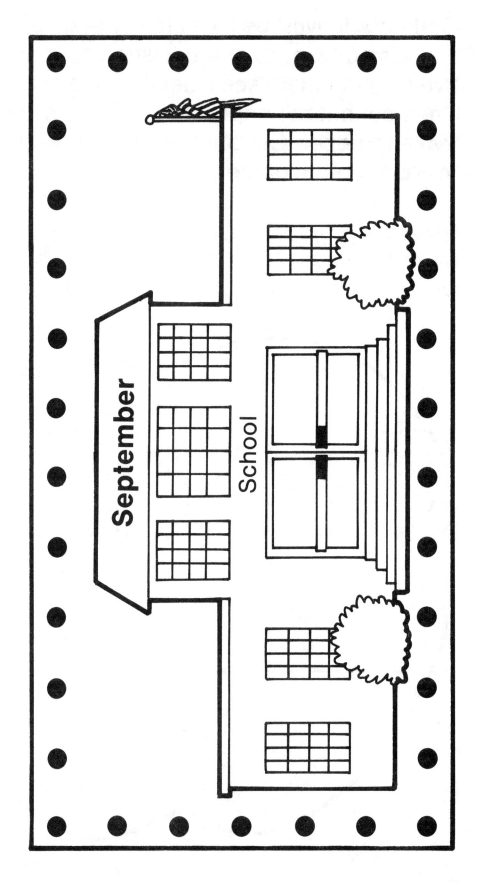

Time: Months, Holidays, Seasons, and Telling Time © 1988

Name _____

Look at the word below.
With crayons, trace each letter.
Make the letter **S** blue.
Make each letter **e** black.
Make the letter **p** yellow.
Make the letter **t** orange.
Make the letter **m** green.
Make the letter **b** red.
Make the letter **r** brown.
Draw red apples around the name of the month.

September

September is the ninth month of the year.
Trace the name of this month.

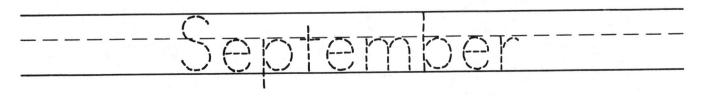

October Calendar

Name _____

Number the days.
Color the picture below.

October

Sunday	Monday	Tuesday	Wednesday	Thursday	Friday	Saturday

Time: Months, Holidays, Seasons, and Telling Time © 1988

October Ghost

Name _____

Put glue along the outside edge of the ghost.
Place dried beans end to end along the glued edge.
Glue on beans for the eyes, the nose, and the mouth.

Time: Months, Holidays, Seasons, and Telling Time © 1988

Name _____

Put glue on the circle. Place small pieces of orange tissue paper over the glue, overlapping the pieces.
Let the glue dry. Then cut out the circle.
Glue the circle to another piece of paper.
Color the cat black. Glue the cat on the moon.
Glue yarn on the cat for whiskers.

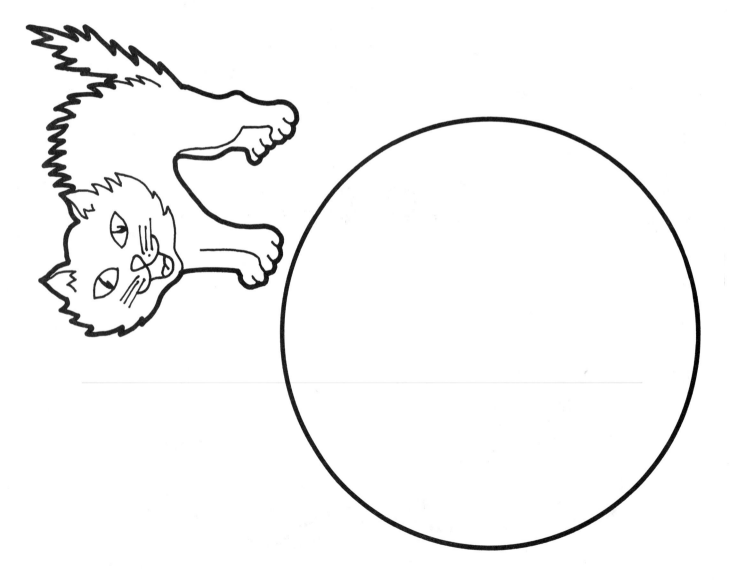

Halloween Bat Lace-up

Name _____

Color the Halloween bat. Glue this page onto tagboard.
Cut out the bat. Punch out a hole at each dot.
Sew around the bat with a lace, and tie the ends.

Writing the Month Name

Name _____

Look at the word below.
With crayons, trace each letter.
Make each letter **o** black.
Make the letter **c** orange.
Make the letter **t** yellow.
Make the letter **b** orange.
Make the letter **e** black.
Make the letter **r** yellow.
Draw orange pumpkins around the name of the month.

October is the tenth month of the year.
Trace the name of this month.

22

Time: Months, Holidays, Seasons, and Telling Time © 1988

November Calendar

Name _____

Number the days.
Color the picture below.

November						
Sunday	Monday	Tuesday	Wednesday	Thursday	Friday	Saturday

Name _____

Thanksgiving Place Setting

Color the plate, the napkin, and the silverware.
Cut out all the pieces of the place setting.
Fringe the opposite edges of a piece of construction paper.
Glue the place setting pieces on the fringed paper, as shown.

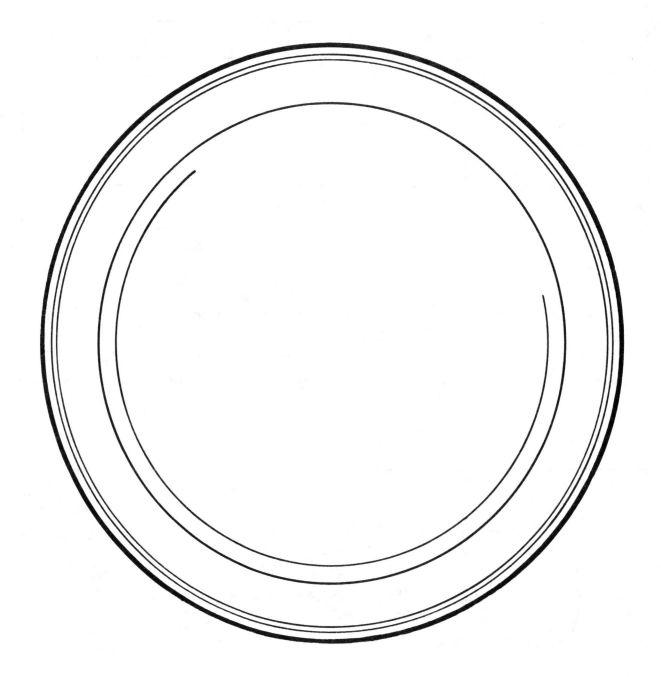

Thanksgiving Place Setting *(continued)*

Name _____

Thanksgiving Turkey Mosaic

Put glue on a small paper plate.
Sprinkle different colors of rice in a pattern on the glue.
Color the turkey's head, body, and feet. Cut them out.
Glue them on the paper plate as shown.

26

Time: Months, Holidays, Seasons, and Telling Time © 1988

Football Season Lace-up

Name _____

Color the football. Glue this page onto tagboard.
Cut out the football. Punch out a hole at each dot.
Sew around the football with a lace, and tie the ends.

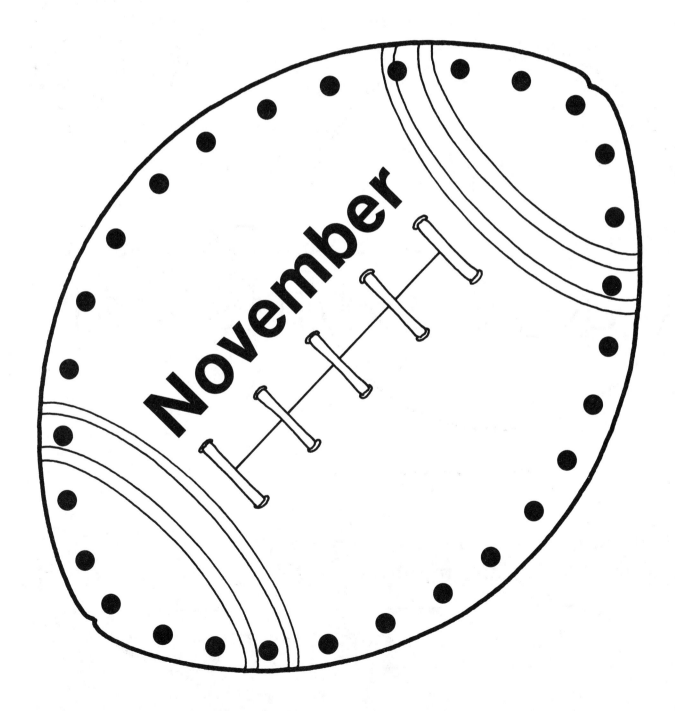

Time: Months, Holidays, Seasons, and Telling Time © 1988

Writing the Month Name

Name _____

Look at the word below.
With crayons, trace each letter.
Make the letter **N** brown.
Make the letters **o**, **m**, and **r** orange.
Make the letters **v** and **b** green.
Make each letter **e** yellow.
Draw brown and yellow leaves around
the name of the month.

November

November is the eleventh month of the year.
Trace the name of this month.

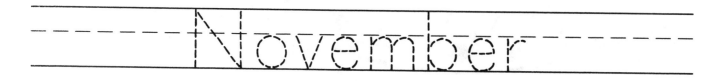

December Calendar

Name _____

Number the days.
Color the picture below.

			December			
Sunday	Monday	Tuesday	Wednesday	Thursday	Friday	Saturday

Time: Months, Holidays, Seasons, and Telling Time © 1988

December Gingerbread

Name _____

Color both sides of the gingerbread boy.
Cut out the gingerbread boy.

December Gingerbread *(continued)*

Staple the body together, leaving the head area open.
Stuff the gingerbread boy with cotton balls.
Staple the head closed.

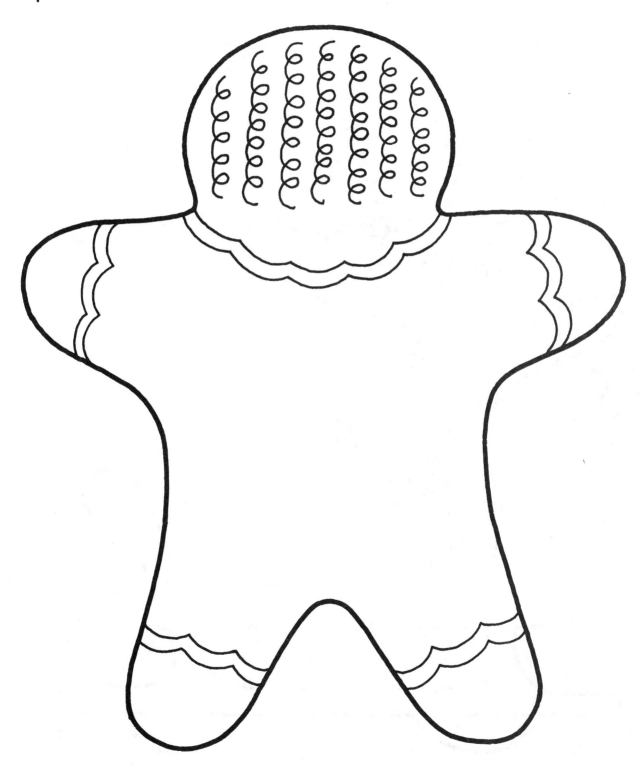

Name _____

December Sleigh Mosaic

Color the sleigh, but do not color the packages.
Draw people inside the sleigh. Cut out pieces of foil paper.
Glue the foil paper over the packages.

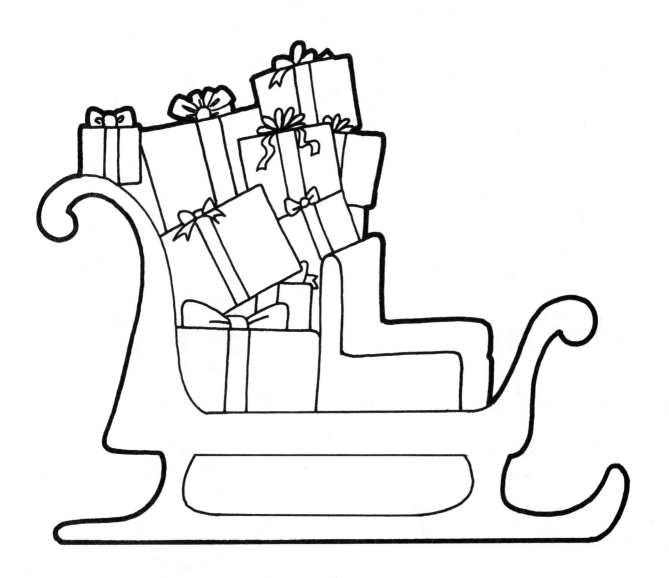

32   Time: Months, Holidays, Seasons, and Telling Time © 1988

Holiday Bell Lace-up

Name _____

Color the holiday bell. Glue this page onto tagboard.
Cut out the bell. Punch out a hole at each dot.
Sew around the bell with a lace, and tie the ends.

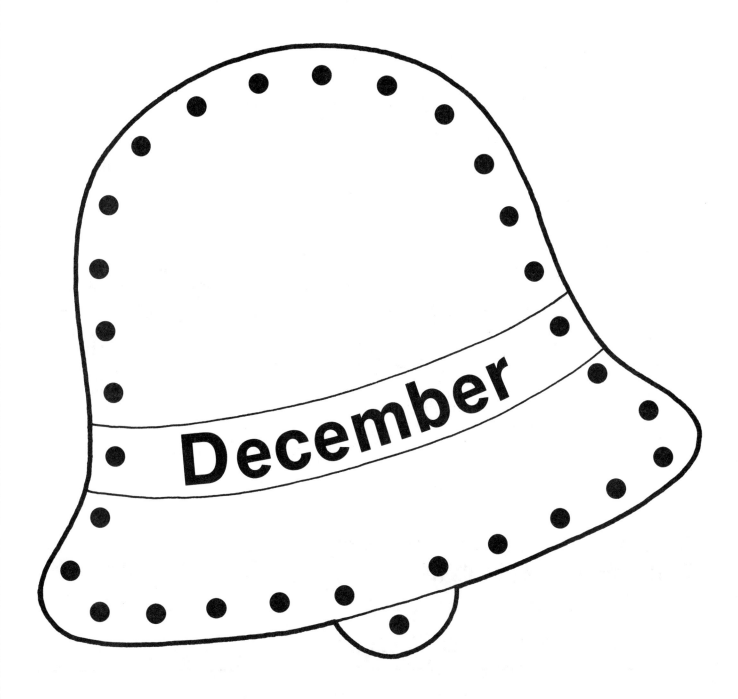

Name _____

Look at the word below.
With crayons, trace each letter.
Make the letter **D** red.
Make each letter **e** green.
Make the letters **c**, **m**, **b**, and **r** red.
Draw red and white striped candy canes around the name of the month.

December

December is the twelfth month of the year.
Trace the name of this month.

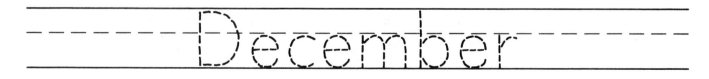

January Calendar

Name _____

Number the days.
Color the picture below.

January

Sunday	Monday	Tuesday	Wednesday	Thursday	Friday	Saturday

Time: Months, Holidays, Seasons, and Telling Time © 1988

New Year's Celebration

Name _____

Color the party decorations.
Write the number that tells what year it is.
Put glue on the letters. Sprinkle glitter on the glue.

January Snowman Mosaic

Name _____

Color the picture background.
Color the snowman's hat and scarf.
Tear white paper into small pieces.
Glue the pieces of torn paper all over the snowman.
Glue on small black circles for eyes and buttons.
Glue on a small orange triangle for the mouth.

Frosty Mitten Lace-up

Name _____

Color the mitten. Glue this page onto tagboard.
Punch out a hole at each dot. Cut out the mitten.
Sew around the mitten with a lace, and tie the ends.

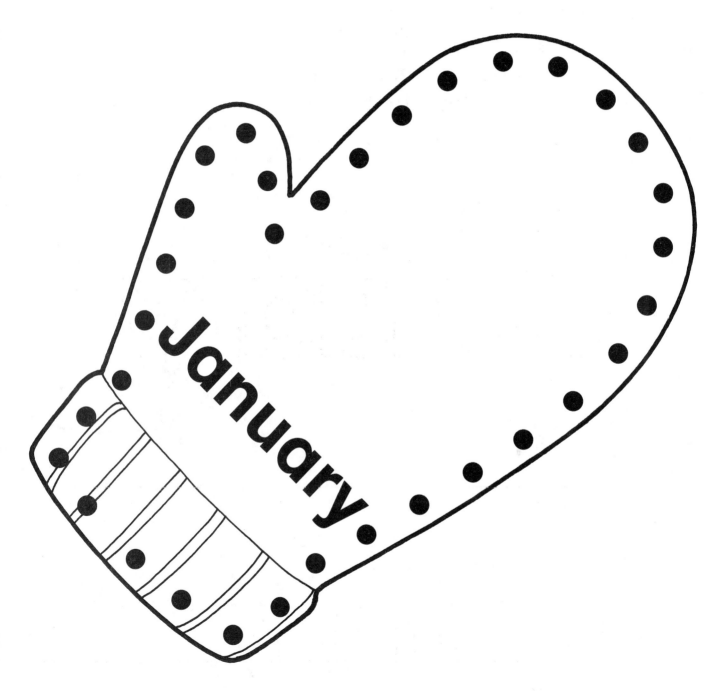

Writing the Month Name

Name _____

Look at the word below.
With crayons, trace each letter.
Make the letter **J** blue.
Make each letter **a** yellow.
Make the letter **n** purple.
Make the letter **u** brown.
Make the letter **r** red.
Make the letter **y** green.
Draw brown and yellow mittens
around the name of the month.

January is the first month of the year.
Trace the name of this month.

February Calendar

Name _____

Number the days.
Color the picture below.

February

Sunday	Monday	Tuesday	Wednesday	Thursday	Friday	Saturday

February Cherry Tree

Name _____

Punch holes from green and brown construction paper.
Gather the hole-punched circles.
Glue brown circles on the outline of the tree trunk.
Glue green circles on the outline of the treetop.
Roll squares of red tissue paper into balls.
Glue the balls on the cherry tree.

Time: Months, Holidays, Seasons, and Telling Time © 1988

Valentine's Day Mosaic

Name _____

Put glue on the big heart. Place small pieces of red tissue paper over the glue, overlapping the pieces. Let the glue dry. Then cut out the heart.
Cut out the small heart. Roll small pieces of white tissue paper in balls. Glue the balls of tissue paper on the small heart. Glue the big heart on another piece of paper.
Glue the small heart on top.

Lincoln Log Cabin Lace-up

Name _____

Color the log cabin. Glue this page onto tagboard.
Cut out the log cabin. Punch out a hole at each dot.
Sew around the log cabin with lace, and tie the ends.

Time: Months, Holidays, Seasons, and Telling Time © 1988

Name _____

Writing the Month Name

Look at the word below.
With crayons, trace each letter.
Make the letters **F**, **b**, and **y** red.
Make each letter **r** black.
Make the letters **e**, **u**, and **a** blue.
Draw red hearts around the name of the month.

February is the second month of the year.
Trace the name of this month.

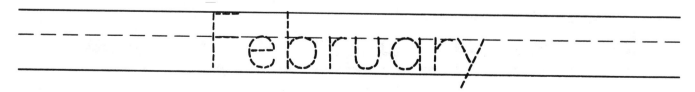

44

Name _____

March Calendar

Number the days.
Color the picture below.

			March			
Sunday	Monday	Tuesday	Wednesday	Thursday	Friday	Saturday

Time: Months, Holidays, Seasons, and Telling Time © 1988

March Lamb

Name _____

Trace the outside edge of the lamb with a black crayon.
Color the lamb's face, ears, and legs pink.
Curl small pieces of yarn around the end of your finger.
Glue the curled yarn to the lamb's furry head.

March Shamrock Mosaic

Name _____

Tear green paper into small pieces.
Glue the torn pieces all over the shamrock.

Time: Months, Holidays, Seasons, and Telling Time © 1988

Leprechaun Hat Lace-up

Name _____

Color the leprechaun's hat. Glue this page onto tagboard.
Cut out the hat. Punch out a hole at each dot.
Sew around the hat with a lace, and tie the ends.

Writing the Month Name

Name _____

Look at the word below.
With crayons, trace each letter.
Make the letters **M** and **h** yellow.
Make the letters **a** and **c** green.
Make the letter **r** orange.
Draw green shamrocks around
the name of the month.

March is the third month of the year.
Trace the name of this month.

49

Time: Months, Holidays, Seasons, and Telling Time © 1988

Name _____

April Calendar

Number the days.
Color the picture below.

April

Sunday	Monday	Tuesday	Wednesday	Thursday	Friday	Saturday

50

Time: Months, Holidays, Seasons, and Telling Time © 1988

Name _____

April Rains

Color the umbrella, the boots, the pant legs, and the raindrops. Cut out each item.
Glue the items on another piece of paper to make this picture.

Time: Months, Holidays, Seasons, and Telling Time © 1988

Easter Egg Mosaic

Name _____

Color the wavy stripes and zigzag stripes.
Glue decorations on the eggs.
Fringe a strip of green construction paper for grass.
Glue the grass strip across the bottom of the page.

Cottontail Lace-up

Name _____

Color the rabbit. Glue it onto tagboard.
Cut out the rabbit. Punch out a hole at each dot.
Sew around the rabbit with a lace, and tie the ends.

April

Time: Months, Holidays, Seasons, and Telling Time © 1988

Writing the Month Name

Name _____

Look at the word below.
With crayons, trace each letter.
Make the letter **A** blue.
Make the letter **p** yellow.
Make the letter **r** red.
Make the letter **i** orange.
Make the letter **l** brown.
Draw green and orange kites
around the name of the month.

April is the fourth month of the year.
Trace the name of this month.

May Calendar

Name _____

Number the days.
Color the picture below.

Sunday	Monday	Tuesday	Wednesday	Thursday	Friday	Saturday
			May			

Children Around the Maypole

Name _____

Color the children and the maypole. Cut them out.
Glue the maypole on a piece of construction paper.
Glue the children around the maypole.
Glue yarn from each child to the maypole.

56 Time: Months, Holidays, Seasons, and Telling Time © 1988

Mother's Day Mosaic

Name _____

Trace the Mother's Day message.
Put glue on the meadow.
Place small pieces of green tissue paper over the glue.
Let the glue dry.
Cut out small flowers from colored paper.
Glue the flowers on the green meadow.
Glue a dried pea at the center of each flower.
Cut along the line. Glue the picture on construction paper.

Time: Months, Holidays, Seasons, and Telling Time © 1988

May Flower Lace-up

Name _____

Color the flower. Glue this page onto tagboard.
Cut out the flower. Punch out a hole at each dot.
Sew around the flower with a lace, and tie the ends.

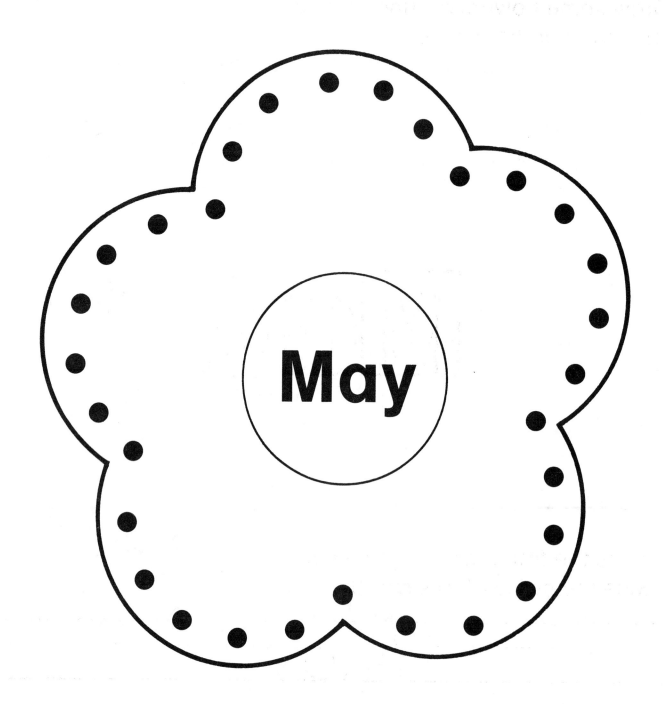

Writing the Month Name

Name _____

Look at the word below.
With crayons, trace each letter.
Make the letter **M** red.
Make the letter **a** yellow.
Make the letter **y** blue.
Draw some flowers around
the name of the month.

May is the fifth month of the year.
Trace the name of this month.

Name _____

June Calendar

Number the days.
Color the picture below.

June						
Sunday	Monday	Tuesday	Wednesday	Thursday	Friday	Saturday

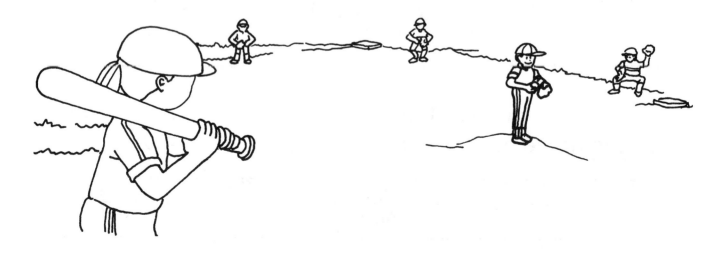

Time: Months, Holidays, Seasons, and Telling Time © 1988

Father's Day Card

Name _____

Color the picture. Cut along the line.
Trace the Father's Day message.
Glue the picture on construction paper.

Happy Father's Day

Time: Months, Holidays, Seasons, and Telling Time © 1988

61

Name _____

Put glue on a piece of blank paper.
Place small pieces of brown tissue paper over the glue, overlapping the pieces.
Let the glue dry. Then trim the edge of the paper.
Color the playground equipment below.
Cut out the pieces and glue them on the brown background.

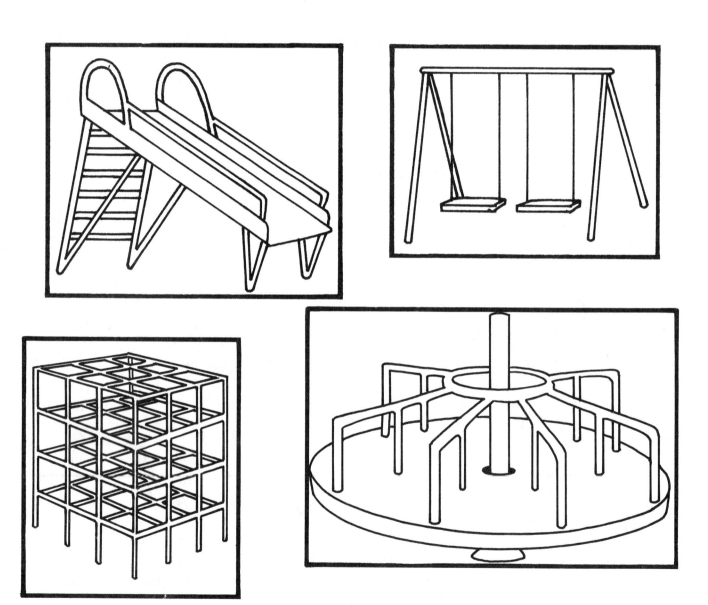

Vacation Travel Lace-up

Name _____

Color the picture. Glue this page onto tagboard.
Cut along the line. Punch out a hole at each dot.
Sew around the picture with a lace, and tie the ends.

June

Time: Months, Holidays, Seasons, and Telling Time © 1988

Name _____

Look at the word below.
With crayons, trace each letter.
Make the letter **J** green.
Make the letter **u** brown.
Make the letter **n** yellow.
Make the letter **e** blue.
Draw purple and yellow butterflies
around the name of the month.

June is the sixth month of the year.
Trace the name of this month.

Name _____

July Calendar

Number the days.
Color the picture below.

July

Sunday	Monday	Tuesday	Wednesday	Thursday	Friday	Saturday

Time: Months, Holidays, Seasons, and Telling Time © 1988

July Picnic

Name _____

Color the picnic basket. Color the food items below. Cut apart the food items. Glue the food inside the basket.

66   Time: Months, Holidays, Seasons, and Telling Time © 1988

Fourth of July Mosaic

Name _____

Color the stripes of the flag white and red.
Tear blue construction paper into small pieces.
Glue the pieces on the flag square, overlapping the pieces.
Put 13 star stickers on the blue square.
The first flag of the United States had only 13 stars.
Now it has 50 stars.

Sailboat Lace-up

Name _____

Color the sailboat. Glue this page onto tagboard.
Cut out the sailboat. Punch out a hole at each dot.
Sew around the sailboat with a lace, and tie the ends.

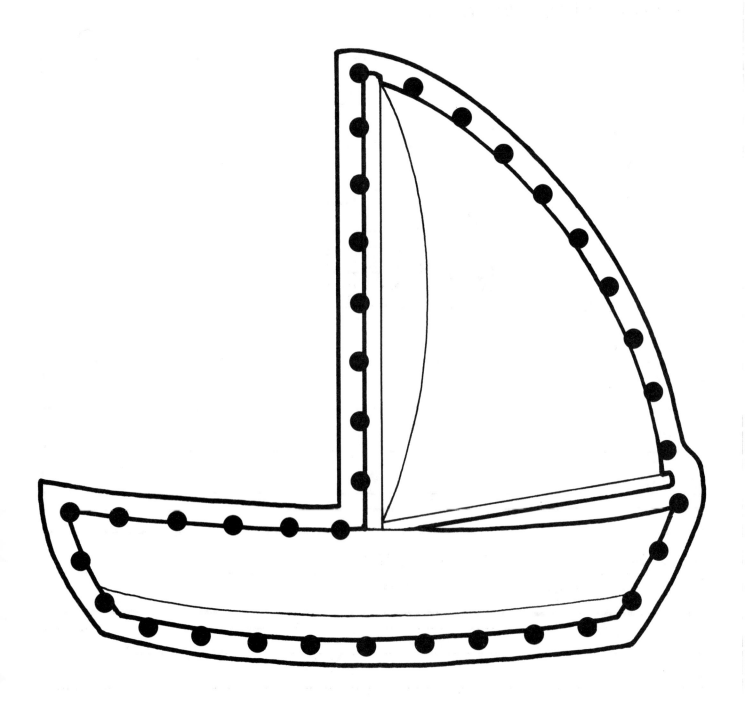

Name _____

Look at the word below.
With crayons, trace each letter.
Make the letters **J** and **l** red.
Make the letters **u** and **y** blue.
Draw red and blue stars around
the name of the month.

July is the seventh month of the year.
Trace the name of this month.

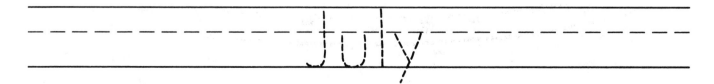

August Calendar

Name _____

Number the days.
Color the picture below.

			August			
Sunday	Monday	Tuesday	Wednesday	Thursday	Friday	Saturday

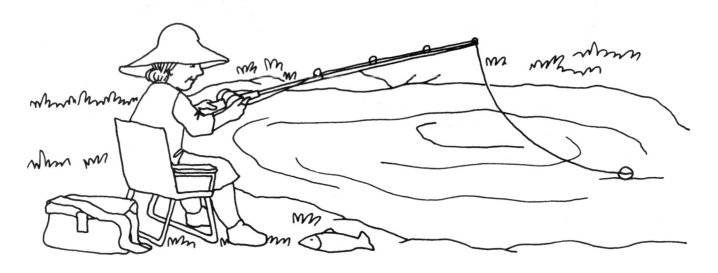

70 Time: Months, Holidays, Seasons, and Telling Time © 1988

August Beach Things

Name _____

Color the beach toys.
Draw more beach toys. Color them.
Put glue across the bottom of the page.
Sprinkle sand on the glue.

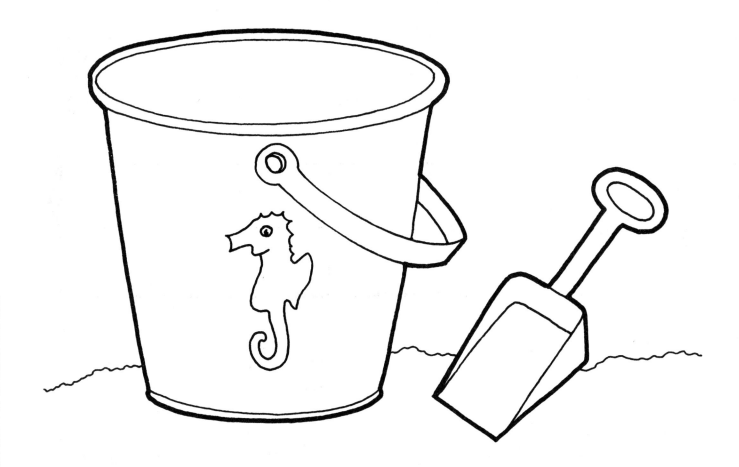

Time: Months, Holidays, Seasons, and Telling Time © 1988

Name _____

August School Box Mosaic

Tear construction paper into small pieces.
Glue the pieces on the school box, overlapping the pieces.
Cut out pictures of school supplies from old magazines and newspapers.
Glue them in place in the school box.

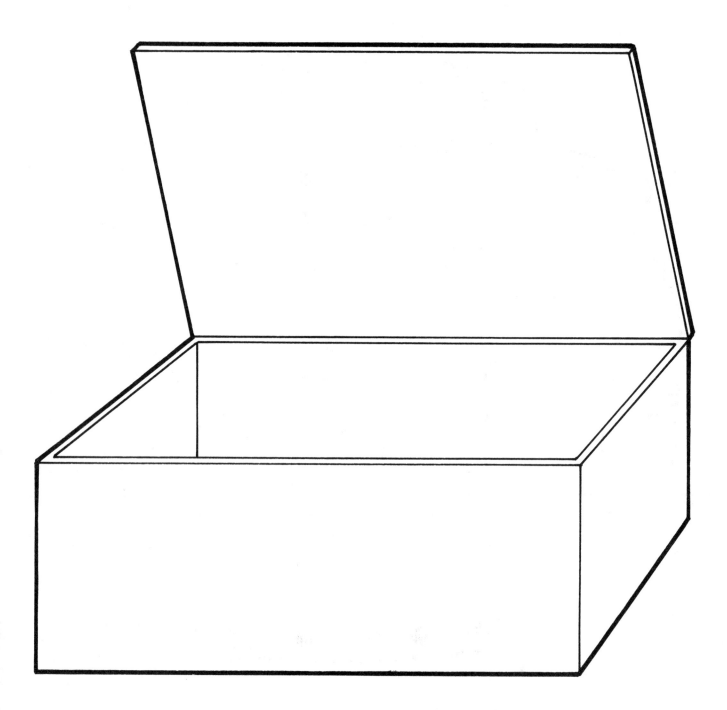

Beach Shell Lace-up

Name _____

Color the shell. Glue this page onto tagboard.
Cut out the shell. Punch out a hole at each dot.
Sew around the shell with a lace, and tie the ends.

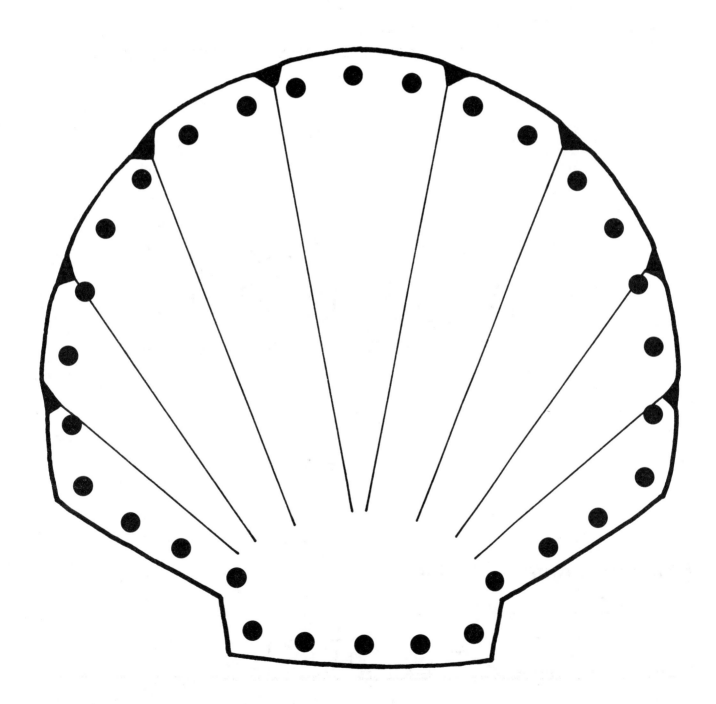

Time: Months, Holidays, Seasons, and Telling Time © 1988

Writing the Month Name

Name _____

Look at the word below.
With crayons, trace each letter.
Make the letters **A** and **t** yellow.
Make each letter **u** orange.
Make the letters **g** and **s** green.
Draw blue and yellow beach balls
around the name of the month.

August is the eighth month of the year.
Trace the name of this month.

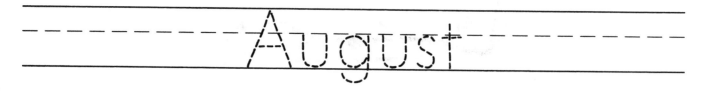

74 Time: Months, Holidays, Seasons, and Telling Time © 1988

Fall Circles

Name _____

In the fall, squirrels collect acorns.
Trace the name of the season.
Another name for **fall** is **autumn**.

Punch holes in brown construction paper.
Gather the hole-punched circles.
Glue the circles around the acorn shape.

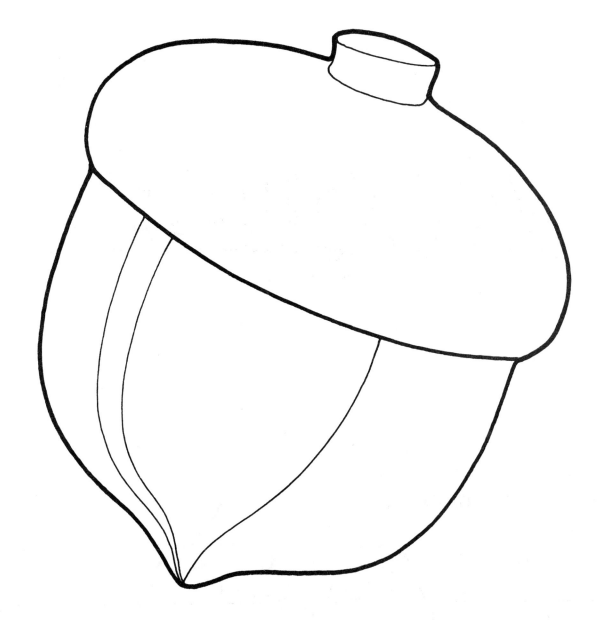

Fall Sponge Painting

Name _____

Color the tree trunk. Dip a sponge into paint.
Dab the sponge on the paper to make leaves.
Make red, yellow, and orange leaves.
Use a different sponge for each color of paint.

Fall Scene

Name _____

Color the fall scene, using bright crayons. Press heavily.
Paint over the scene with a light-colored paint.
Use left-to-right brushstrokes.

Time: Months, Holidays, Seasons, and Telling Time © 1988

Winter Circles

Name _____

In most places, it snows in the winter.
Every snowflake has six main points.
Trace the name of the season.

Punch holes in light-blue construction paper.
Gather the hole-punched circles.
Glue the circles on the lines of the snowflake.

Winter Sponge Painting

Name _____

Trace the tree trunks and branches with a brown crayon.
Dip a sponge into green paint.
Dab the sponge on the paper to make pine needles.

Winter Scene

Name _____

Color the winter scene, using bright crayons. Press heavily.
Paint over the scene with a light-colored paint.
Use left-to-right brushstrokes.

80

Time: Months, Holidays, Seasons, and Telling Time © 1988

Spring Circles

Name _____

Spring is the season for rain.
Trace the name of this season.

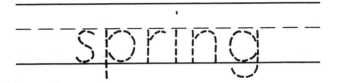

Color the picture, but do not color the sky.
Punch holes in gray construction paper.
Gather the hole-punched circles.
Glue the circles in the sky for raindrops.

Time: Months, Holidays, Seasons, and Telling Time © 1988

Spring Sponge Painting

Name _____

Color the leaves and stems of the daffodils.
Dip a sponge into yellow paint.
Dab the sponge on the paper to make the daffodil petals.

Spring Scene

Name _____

Color the spring scene, using bright crayons. Press heavily.
Paint over the scene with a light-colored paint.
Use left-to-right brushstrokes.

Time: Months, Holidays, Seasons, and Telling Time © 1988

Summer Circles

Name _____

Summer is the hottest month.
The sun shines a long time
each day.
Trace the name of this season.

Punch holes in orange and yellow construction paper.
Gather the hole-punched circles.
Glue the circles on the outline of the sun.

Summer Sponge Painting

Name _____

Color the fence around the rose bushes.
Dip a sponge into paint.
Dab the sponge on the paper to make roses.
Make red, pink, yellow, and purple roses.
Paint the leaves green.
Use a different sponge for each color of paint.

Summer Scene

Name _____

Color the summer scene, using bright crayons. Press heavily.
Paint over the scene with a light-colored paint.
Use left-to-right brushstrokes.

Reviewing the Seasons

Name _____

Draw a line to match each season with a picture.
Color the pictures.

Fall

Winter

Spring

Summer

Time: Months, Holidays, Seasons, and Telling Time © 1988

Reviewing the Seasons

Name _____

Write the name of the season for each picture.
Color the pictures.

_____

_____

_____

_____

Clock Art

Name _____

Complete the picture below.
Color the picture.

Name _____

Hidden Time

Find six clocks hidden in the picture below.
Circle each one.

90

Time: Months, Holidays, Seasons, and Telling Time © 1988

Name _____

Clock Numbers

Look at each clock.
Fill in the missing numbers.

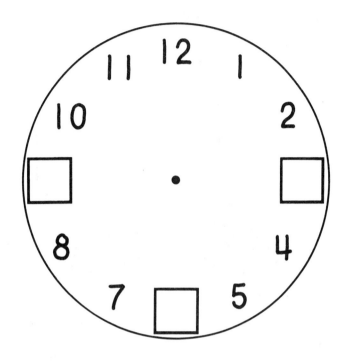

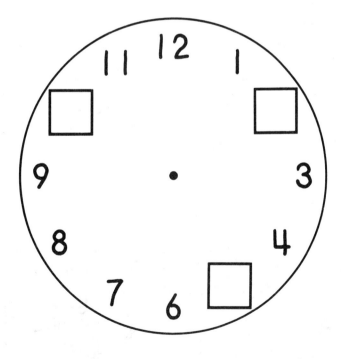

Clock Numbers

Name _____

Look at each clock.
Fill in the missing numbers.

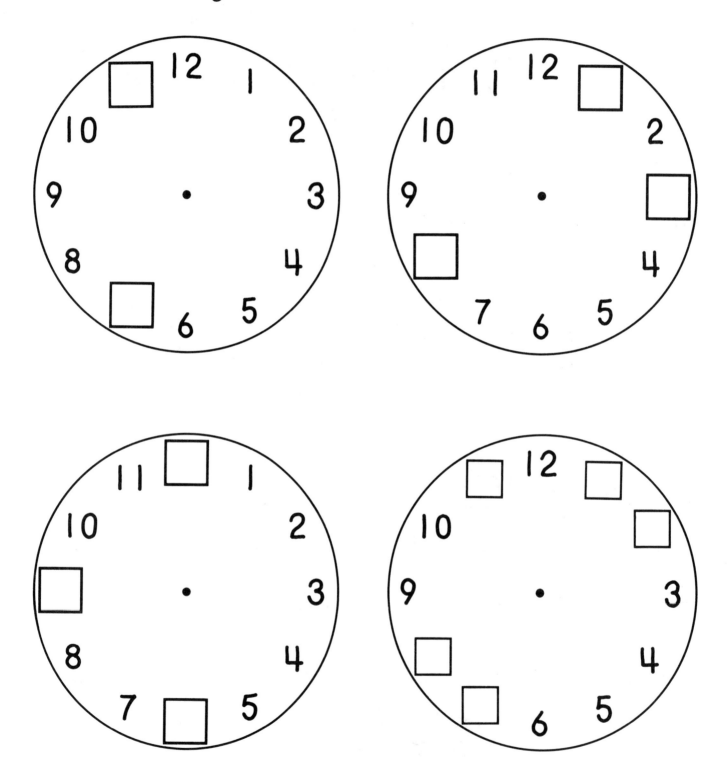

Telling Time

Name _____

The clock shows 3 o'clock.
Trace the time.

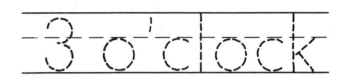

Look at the little hand on the clock.
Write the number it points to.

_____

- - - - - - - - - - - - - - -

_____

Show 3 o'clock.
Draw the clock hands.

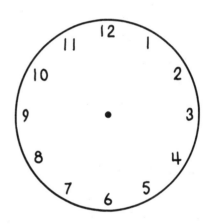

At 3 o'clock, we play outdoors.
Color the picture.

Time: Months, Holidays, Seasons, and Telling Time © 1988

Telling Time

Name _____

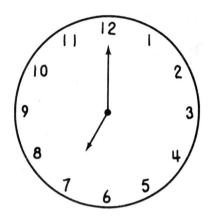

The clock shows 7 o'clock.
Trace the time.

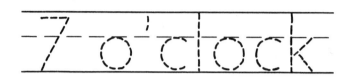

Look at the little hand on the clock.
Write the number it points to.

Show 7 o'clock.
Draw the clock hands.

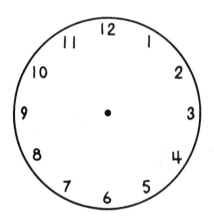

At 7 o'clock in the morning, we wake up.
Color the picture.

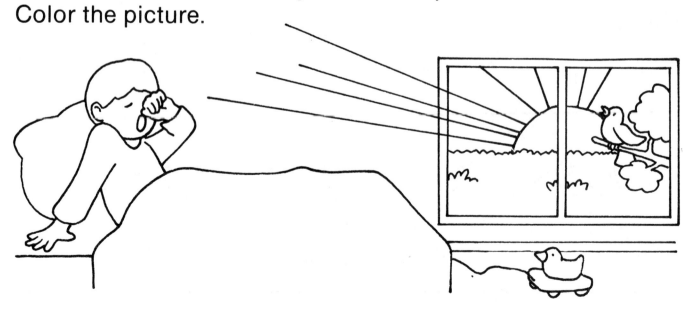

94
Time: Months, Holidays, Seasons, and Telling Time © 1988

Daytime

Name _____

What do you see in the sky during the day?
Color each thing you might see.

How does the sky look during the day?
Trace the word.

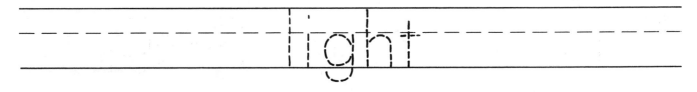

Nighttime

Name _____

What do you see in the sky at night?
Color each thing you might see.

How does the sky look at night?
Trace the word.